HIBERNATION

NATURE'S CYCLES

Mel Higginson

Rourke
Publishing LLC
Vero Beach, Florida 32964

www.rourkepublishing.com

PHOTO CREDITS: All photographs © Lynn M. Stone except p. 5 © www.Sweetmoments.pl, P. 22 © Swopedesign

Editor: Robert Stengard-Olliges

Cover and interior design by Nicola Stratford

Library of Congress Cataloging-in-Publication Data

Higginson, Mel.
   Hibernation / Mel Higginson.
      p. cm. -- (Nature's cycle)
   Includes bibliographical references (p. 24).
   ISBN 1-60044-177-7 (hardcover)
   ISBN 1-59515-535-X (softcover)
   1. Hibernation--Juvenile literature. I. Title. II. Series: Higginson, Mel. Nature's cycle.
   QL755.S76 2007
   591.56'5--dc22   2006013606

Printed in the USA

CG/CG

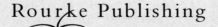

Rourke Publishing

www.rourkepublishing.com – sales@rourkepublishing.com
Post Office Box 3328, Vero Beach, FL 32964

Table of Contents

Hibernation

Hibernation is a deep sleep. It is a way for some kinds of animals to **survive** winter.

Winter cold and snow make food hard to find. But animals in hibernation do not have to find food!

Animals That Hibernate

Bats live on insects. But insects disappear in winter.

Bats solve their food problem by hibernating. They enter **caves** and fall into a deep sleep.

Marmots are plump ground squirrels. Marmots live on mountains. Their little gardens disappear in winter.

Marmots cannot fly away from winter like geese do. So…they **hibernate**! In fact, marmots spend most of the year in hibernation!

Grizzly bears eat plants and other animals. Their food disappears in winter.

Snakes, turtles, frogs, and toads living in cold places hibernate. Very few birds hibernate. Most birds **migrate** away from winter.

Furry animals **prepare** for hibernation by eating a lot. Their bodies store more fat.

Like marmots, grizzlies fatten up in the summer and fall. Then they curl up and hibernate.

Body Changes for Hibernation

During hibernation, animals live in slow motion. Their heart slows down. Their breathing slows down. Their body temperature drops.

Food is fuel for active animals. Hibernating animals are not active. What little food they need comes from body fat.

Some kinds of hibernating animals can wake up at any time. They spend most of the winter asleep, however.

The warm weather of spring stirs plants to new growth. It also stirs hibernating animals to wake up!

Glossary

caves (KAYVZ) — large holes underground
hibernate (HYE bur nate) — a deep sleep during winter
migrate (MYE grate) — to move from one place
 to another
prepare (pri PAIR) — to get ready
survive (sur VIVE) — to stay alive

INDEX

FURTHER READING

Ganeri, Anita. *Hibernation*. Heinemann, 2005.

Malcolm, Penny. *Hidden Hibernators*. Heinemann, 2004.

WEBSITES TO VISIT

http://www.sciencemadesimple.com/animals.html

http://dnr.wi.gov/org/caer/ce/eek/nature/snugsnow.htm

ABOUT THE AUTHOR

Mel Higginson writes children's nonfiction and poetry. This is Mel's first year writing for Rourke Publishing. Mel lives with his family just outside of Tucson, Arizona.